MINAHIL MUNAWAR
KAINAT ABBAS
ALIM UN NISA

HUMAN MICROBIOME

MINAHIL MUNAWAR
KAINAT ABBAS
ALIM UN NISA

HUMAN MICROBIOME

We Are Not Alone: The Hidden World within Us

Noor Publishing

Imprint
Any brand names and product names mentioned in this book are subject to trademark, brand or patent protection and are trademarks or registered trademarks of their respective holders. The use of brand names, product names, common names, trade names, product descriptions etc. even without a particular marking in this work is in no way to be construed to mean that such names may be regarded as unrestricted in respect of trademark and brand protection legislation and could thus be used by anyone.

Cover image: www.ingimage.com

Publisher:
Noor Publishing
is a trademark of
Dodo Books Indian Ocean Ltd. and OmniScriptum S.R.L publishing group

120 High Road, East Finchley, London, N2 9ED, United Kingdom
Str. Armeneasca 28/1, office 1, Chisinau MD-2012, Republic of Moldova, Europe
Managing Directors: Ieva Konstantinova, Victoria Ursu
info@omniscriptum.com

Printed at: see last page
ISBN: 978-620-8-87100-0

HUMAN MICROBIOME

We Are Not Alone: The Hidden World within Us

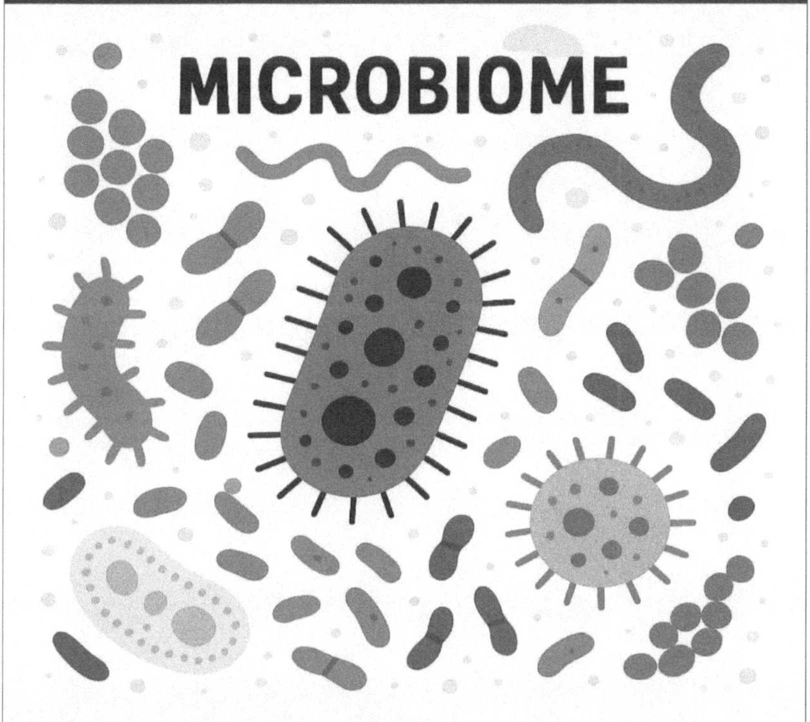

MINAHIL MUNAWAR
KAINAT ABBAS
ALIM UN NISA

Dedication

To my beloved parents, Father **Munawar Hussain** and my Mother **Uzma Bashir** whose endless love, sacrifices, and prayers have shaped the person I am today. And to **Shafqat Malik** the one man who quietly lives in the deepest corner of my heart you are my calm in the storm, my truth in a noisy world. Your unwavering belief in me has lit the darkest paths. With you, I have found a love that is rare, real, and forever. This book is a piece of my soul and so much of it is shaped by you.

Minahil Hussain

Table of Contents

Table of Figure

Chapter No 01

Introduction to the Human Microbiome

What is Microbiome?

- Micro From Greek mikros, meaning small or tiny
- Bio From Greek bios, meaning life
- Ome From Greek oma, used in science to refer to a complete system or set

Definition:

The human microbiome is the sum total of all microorganisms (fungi, viruses, bacteria, and other microbes) that inhabit and reside on the human body. These microorganisms mostly live in places such as the skin, gut, mouth, and other body systems. The human gut microbiome, specifically, harbors trillions of microorganisms and plays vital functions in the following terms;

⇒ Digestion and metabolism
⇒ Immune system development and function
⇒ Protection against pathogens
⇒ Production of essential vitamins and compounds
⇒ Influence on mental health through the gut-brain axis

Types of Human Microbiome?

There are some types of microorganisms that build up the human microbiome. Here's a detailed of the major categories:

- **Types of Microorganisms in the Human Microbiome**

1. Bacteria

Bacteria form the highest percentage of the human microbiome. Bacteria are one-celled, non-nuclear prokaryotes. Most significant bacterial genera within the human microbiome include:

- Firmicutes: Gram-positive bacteria such as *Lactobacillus, Bacillus,* and *Clostridium*. They form a high percentage in the gut microbiome and partake in the metabolism of carbohydrate and energy gathering.

- Bacteroidetes: Gram-negative bacteria that specialize in the gut's intestinal tract. They focus on degrading large polysaccharides as well as plant matter.

- Actinobacteria: Contains useful genera such as *Bifidobacterium* which are essential for infant gut health and immune response.

- Proteobacteria: A heterogeneous phylum containing both useful species and possible pathogens such as *E. coli.* Typically found at low abundance in healthy individuals.

- Verrucomicrobia: Contains species such as *Akkermansia muciniphila* that helps maintain mucus layer and metabolic health.

2. Archaea

Archaea are unicellular microbes that, similar to bacteria, do not have a nucleus but are distinct in cell membrane composition and metabolism. Within the human microbiome:

- Methanobrevibacter: Most prevalent archaeal genus in humans, specifically *M. smithii,* which assists in fermenting otherwise unusable polysaccharides and eliminating excess hydrogen through methane production.

- Nitrososphaera: Present on the skin, with possible roles in nitrogen cycling.

3. Fungi (Mycobiome)

Fungi make up a lesser but not insignificant part of the human microbiome:

- Candida: Multiple species colonize various body sites such as the gut, skin, and vagina. *C. albicans* is the most prevalent species.

- Malassezia: Predominant on human skin, especially in sebum-covered areas. They're linked to diseases such as dandruff and *seborrheic dermatitis.*

- Saccharomyces: Such as *S. cerevisiae* (baker's yeast), which can transiently colonize the gut.

- Aspergillus: Typically environmental fungi that can transiently colonize the respiratory tract.

4. Viruses (Virome)

The viral part of the microbiome consists of:

- Bacteriophages: Bacterial viruses, they are the most numerous entities of the human microbiome and play a role in controlling bacterial populations. Everybody site possesses unique phage communities.

- Eukaryotic viruses: Comprise pathogenic viruses and commensal/latent viruses such as herpesviruses that can exist without causing infection.

- Endogenous retroviruses: Retroviral elements integrated into the human genome during evolutionary time, some of which are still transcriptionally active.

5. Protozoa
These are eukaryotic single cells that could be components of the microbiome:

- Blastocystis: Ubiquitous intestinal protist whose health effects are unknown can be commensal in one person and pathogenic in another.

- Entamoeba: Both pathogenic and non-pathogenic species are included that can colonize the gut.

6. Helminths
Although classically thought of as parasites, there is increasing evidence that some helminth (worm) species could have coevolved with humans and have immunoregulatory functions:

- Necator and Trichuris: Low-grade colonization in certain traditional communities may assist in controlling immune function.

These various types of microorganisms do not exist alone but create intricate ecological communities, with their interactions being competition, cooperation, predation, and metabolic interchange. This variety is responsible for the human microbiome's excellent functional capacity and its influence on health.

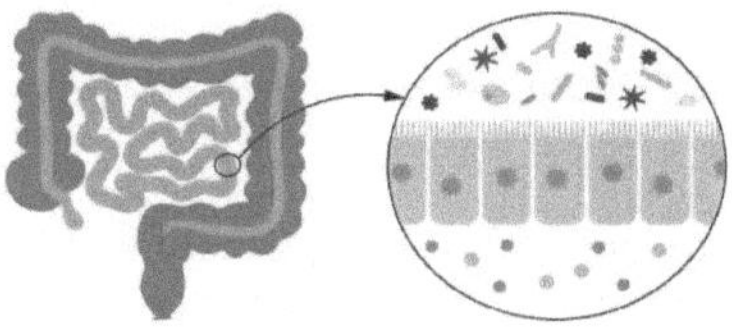

Figure 1: Shows the human gut and microbiome

Microbiome vs. Microbiota

Microbiota:

Definition:

Microbiota refers specifically to the living microbial organisms themselves the collection of bacteria, archaea, fungi, protists, and viruses that inhabit a particular environment.

Characteristics:

- Represents only the microorganisms (the biological entities)
- Describes the taxonomic composition of microorganisms present

- Can be analyzed through methods like 16S rRNA sequencing to identify which species are present
- Focuses on "who is there" in a microbial community
- The term emphasizes the ecological community of commensal, symbiotic, and pathogenic microorganisms

Example:

The gut microbiota would include specific bacterial species like *Lactobacillus acidophilus, Bifidobacterium longum,* and others actually present in the intestinal tract.

Microbiome:

Definition:

Microbiome encompasses both the microorganisms (microbiota) and their collective genomes, including all their genes and gene products, as well as the surrounding environmental conditions.

Characteristics:

- More comprehensive, combining both the microorganisms and their genetic material
- Includes the functional capacity encoded in these genomes
- Analyzes what these microorganisms can potentially do based on their genetic blueprints
- Focuses on both "who is there" and "what are they capable of doing"
- Often studied through metagenomic, metatranscriptomic, and metabolomic approaches
- Includes the microbes' interactions with their environment (host tissue, nutrients, etc.)

Example:

The gut microbiome would include not just the microbial species present, but also the millions of genes they contain, the metabolites they produce, and how these interact with the gut environment.

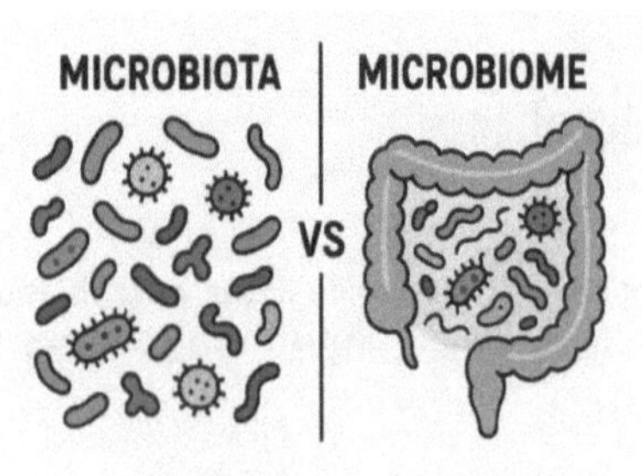

Figure 2: Shows the difference in microbiota and microbiome

Overview of Major Body Sites Colonized

✓ **Gut Microbiome**

The biggest and most varied microbiome in the human body resides in the gastrointestinal tract. From the mouth all the way to the large intestine, the gut microbiome is home to about 100 trillion microbes. These microbe colonizers contribute to digestion, synthesize vital vitamins (specifically B vitamins and vitamin K), modulate immune function, and affect metabolism. The microbial density differs along the digestive tract, and the greatest microbial population is found in the colon. Major bacterial phyla are *Firmicutes, Bacteroidetes, Actinobacteria*, and *Proteobacteria*, although the particular composition varies among individuals depending on issues such as diet, genetics, and environment.

✓ **Skin Microbiome**

The skin microbiome differs considerably along body regions as a result of differences in pH, moisture, temperature, and sebum. Dry regions such as forearms have different populations than wet regions such as underarm or oily areas such as the face. *Staphylococcus, Corynebacterium, Propionibacterium,* and *Malassezia* are common genera. The bacteria create an umbrella against infection-causing microbes, modulate the immunity of the skin, and affect acne, eczema, and psoriasis.

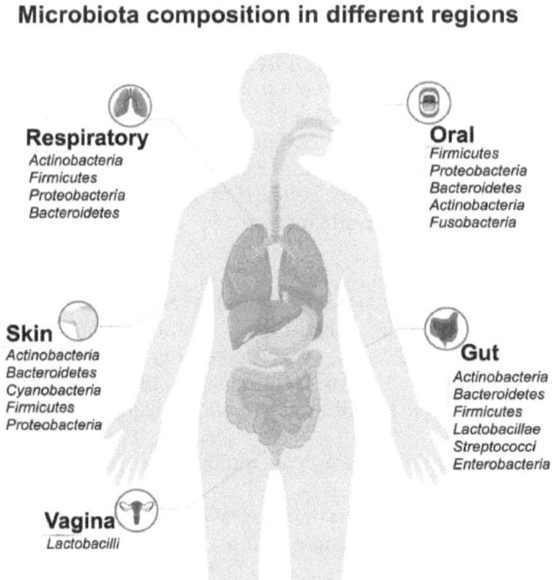

Figure 3: Human Microbiota colonized at different body sites

✓ **Oral Microbiome**

The oral cavity consists of a range of microenvironments such as teeth, gums, tongue, cheeks, and palate, each providing distinct microbial ecosystems. There are more than 700 species of bacteria present in the mouth, some of which major genera like Streptococcus, *Veillonella, Actinomyces,* and *Fusobacterium*. It aids in keeping the oral cavity uncolonized by harmful pathogens but also facilitates disease like dental caries and periodontitis if the flora gets disturbed.

✓ **Respiratory Microbiome**

The respiratory tract, from the nasal cavities to the lungs, is inhabited by its own microbial community. Previously considered sterile, lungs are now understood to carry a low-density but meaningful microbiome. The respiratory tract is richer in microbes at the upper sites, with *Staphylococcus, Propionibacterium,* and *Corynebacterium* species found in the nasal cavity. The respiratory microbiome performs critical functions in immune education and resistance against respiratory infections.

✓ **Urogenital Microbiome**

The female urogenital tract, especially the vagina, is home to a specialized microbiome that is dominated by *Lactobacillus* species that keep the environment acidic and prevent pathogen growth. The male urogenital microbiome has received less attention but consists of communities in the urethra and on external genitalia. This microbiome determines susceptibility to urinary tract infections, sexually transmitted infections, and reproductive health.

✓ **Placental and Breast Milk Microbiome**

Current studies have found distinct microbial communities in the placenta and breast milk, contradicting the long-held belief that these environments are sterile. The microbiome of breast milk includes health-promoting bacteria such as *Bifidobacterium* and *Lactobacillus* that colonize the infant gut and aid in immune development. These microbes are believed to be partially of maternal gut origin and transferred to the infant via specialized routes.

✓ **Ocular Microbiome**

The surface of the eye supports a lean but important microbiome that has evolved for this nutrient-poor, antimicrobial-rich habitat. Representative genera are *Staphylococcus, Streptococcus, Corynebacterium,* and *Propionibacterium.* The microbiome is important to the maintenance of ocular surface integrity and prevention of colonization by potentially pathogenic microorganisms.

Each of these microbiome is in a fragile equilibrium that can be upset by such factors as antibiotics, diet, hygiene habits, and environmental exposure. Contemporary research increasingly appreciates the interrelatedness of these microbial populations and their combined effects on human health.

Your 4 Major MICROBIOME

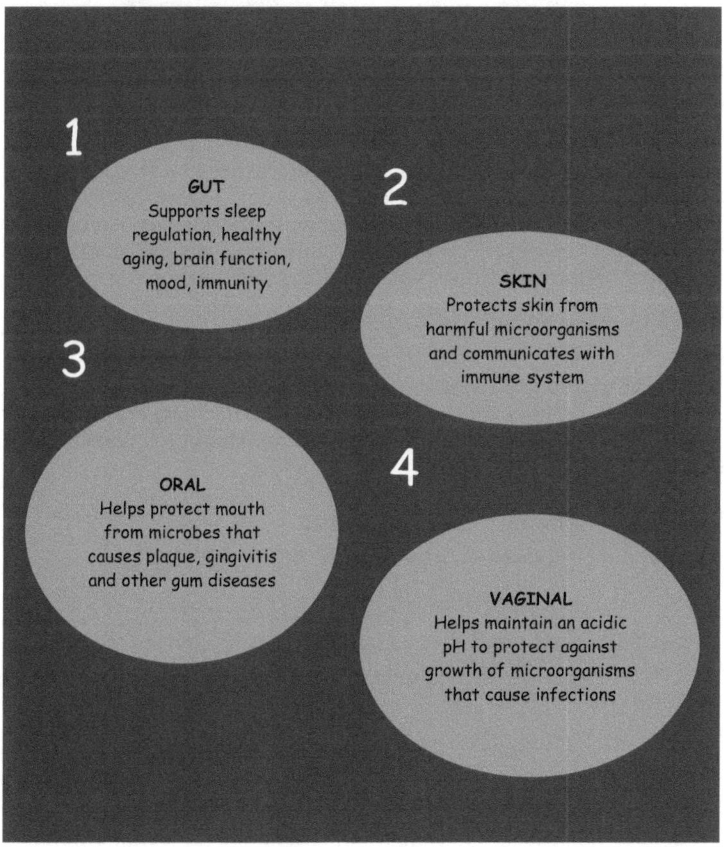

Figure 4: Shows major human body microbiome

Chapter No 02

The Gut Microbiome

What is the Gut Microbiome?

Your "**gut microbiome**" consists of the trillions of microbes and their DNA residing in your intestinal tract. These microbes, which are predominantly bacteria, play roles essential to your health and wellbeing. These bacteria reside within your digestive tract and they have a central function in breaking down food you consume, assisting with nutrient absorption and synthesizing, controlling your metabolism, body weight and immune system, and also assisting in maintaining brain function.

Composition of gut flora and how it varies from person to person

Gut flora, or gut microbiota, is the extensive group of microorganisms living within our digestive tract, particularly in the large intestine. These microbes are incredibly abundant that they actually outnumber human cells and weigh approximately 1–2 kilograms collectively!

How Gut Flora Varies from Person to Person:

No two people have the exact same gut flora. It changes based on:

Diet

- A high-fiber, plant-based diet supports diverse, healthy microbes.
- A Western diet high in sugar, fat, and processed food can reduce good bacteria and increase harmful ones.

Age

- At birth, babies receive their first microbes from the mother (during birth or through breast milk).
- In infancy, *Bifidobacteria* dominate and help build the immune system.
- In adults, microbiota becomes more diverse and stable.
- In elderly people, diversity may decline, and this is linked to poorer health.

Environment

- Growing up in a rural area, playing outside, or living with pets can increase microbial diversity.

- Use of antibiotics, disinfectants, and overly clean lifestyles may decrease good microbes.

Genetics

- Our genes influence which microbes are more likely to live in our gut, though this plays a smaller role than diet and environment.

Health & Disease

- Conditions like obesity, diabetes, allergies, depression, and even autism have been linked to imbalanced gut flora.
- A healthy gut maintains a balance between beneficial and potentially harmful microbes.

Medications

- Antibiotics, while useful, can wipe out both bad and good bacteria.
- Other medications like antacids, painkillers, and antidepressants may also alter gut flora.

How to Maintain Healthy Gut Flora

- Eat fiber-rich foods: fruits, vegetables, whole grains, and legumes
- Include fermented foods: yogurt, kefir, sauerkraut, kimchi
- Limit processed foods and sugar
- Avoid unnecessary antibiotics
- Manage stress: stress can affect gut health
- Stay physically active

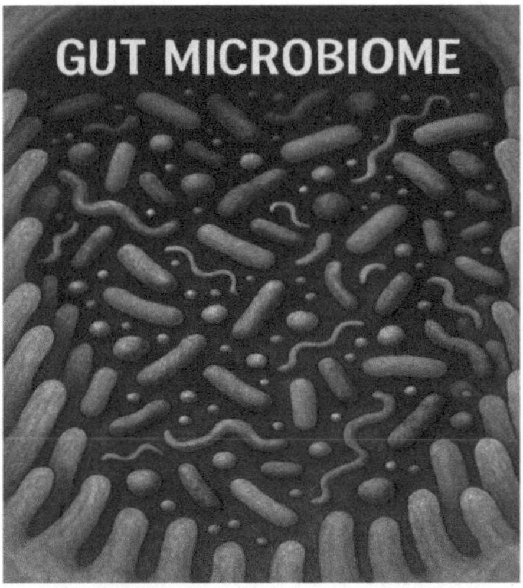

Figure 5: Shows the microbiome of human gut

Role in digestion, metabolism, and vitamin production

The human microbiome, which consists of trillions of microorganisms residing in the gut, plays a crucial role in digestion, metabolism, and vitamin production.

Digestion:

The microbiome helps to process complex carbohydrates and fibers that are not digestible by the human digestive system individually. These bacteria ferment these elements, creating SCFAs which supply energy for colon cells as well as contributing to overall digestive health.

Metabolism:

The gut microbiota controls metabolic functions by impacting the way the body harvests energy from food. It controls fat storage, insulin sensitivity, and metabolism of other nutrients. Some bacteria may also affect calorie and nutrient absorption, thus impacting body weight and metabolic health.

Vitamin Production:

Certain bacteria present in the gut are responsible for producing crucial vitamins, like B vitamins (B12, B6) and vitamin K. These vitamins perform key functions in the body, such as producing energy, blood clotting, and the health of nerve function. As a whole, the human microbiome is critical to keeping the digestive system healthy, controlling metabolism, and aiding in vitamin production, which indicates its significance in overall health.

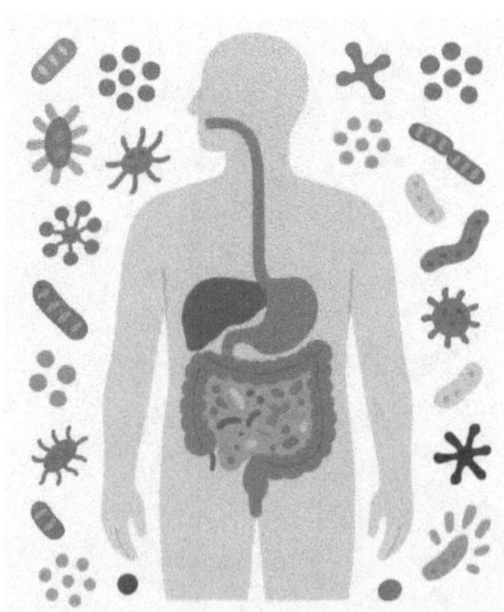

Figure 6: Shows its role in digestion

Link with obesity, diabetes, and irritable bowel syndrome

The human microbiome is tightly connected with obesity, diabetes, and irritable bowel syndrome (IBS) as it controls the metabolism and the health of the gut. An imbalance in the gut bacteria can result in more fat deposition, insulin resistance, and gastrointestinal illness, thereby marking the connection between the microbiome and the illnesses.

Obesity and the Human Microbiome

- **Metabolic Influence:** The gut microbiota affects energy extraction from food, which can lead to increased fat storage. Certain bacterial populations are associated with obesity, as they may promote the fermentation of dietary fibers into short-chain fatty acids (SCFAs), influencing fat accumulation.
- **Inflammation:** Dysbiosis can trigger low-grade inflammation, which is linked to obesity and metabolic syndrome. This inflammation can disrupt insulin signaling, contributing to insulin resistance.

Diabetes and the Human Microbiome

- **Insulin Resistance:** The composition of gut microbiota can influence insulin sensitivity. An imbalance in gut bacteria may lead to increased levels of inflammatory markers, which are associated with type 2 diabetes.
- **Metabolic Pathways:** Specific gut bacteria are involved in the metabolism of carbohydrates and lipids, affecting glucose homeostasis. Changes in the microbiome can alter the metabolism of bile acids, which play a role in glucose metabolism.

Irritable Bowel Syndrome (IBS) and the Human Microbiome

- **Gut Health:** Dysbiosis has been implicated in the pathogenesis of IBS. An imbalance in gut bacteria can lead to symptoms such as bloating, diarrhea, and constipation.
- **Gut-Brain Axis:** The microbiome communicates with the central nervous system, influencing gut motility and sensitivity. This interaction can exacerbate IBS symptoms, highlighting the importance of a balanced microbiome for gut health.

In short the human microbiome plays a significant role in the development and management of obesity, diabetes, and irritable bowel syndrome through its effects on metabolism, inflammation, and gut health.

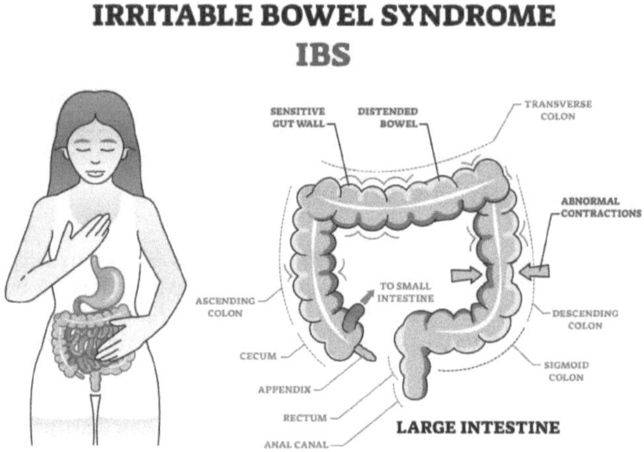

Figure 7: Shows irritable Bowel Syndrome

Concept of leaky gut and gut brain axis

Leaky Gut (Increased Intestinal Permeability)

Leaky gut is a condition where the lining of the small intestine is damaged or compromised, and small gaps are created between the cells. These gaps permit undigested food particles, toxins, and bacteria to pass through into the bloodstream something that shouldn't be allowed.
In a healthy gut, the lining works like a tight filter, passing only nutrients. But when it gets "leaky," the immune system responds to the foreign particles, which can cause inflammation and is believed to be connected with digestive ailments, fatigue, skin issues, allergies, and even autoimmune conditions.

Symptoms of Leaky gut:

- Digestive Problems
- Hormone Problems
- Immunity Problems
- Skin Problems
- Joint Problems
- Metabolism Problems
- Autoimmune Problems
- Mental Health Problem

Possible causes include:

- Poor diet (high in sugar, alcohol, processed foods)

- Chronic stress
- Infections
- Overuse of antibiotics or painkillers

Supporting gut health with a balanced diet, probiotics, and stress management may help improve the gut lining.

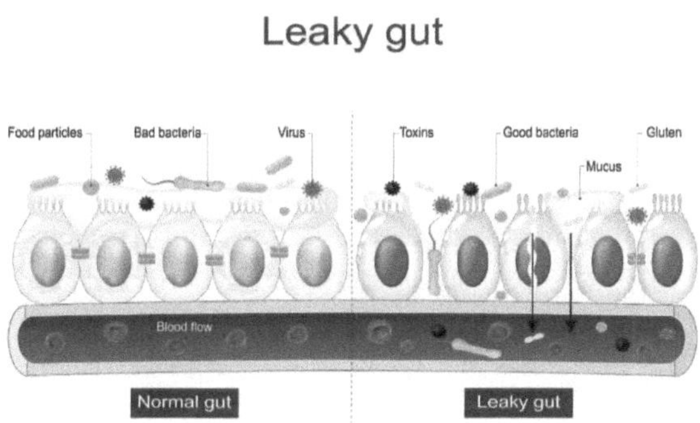

Figure 8: Shows normal and leaky gut

Gut-Brain Axis

The gut-brain axis is the bidirectional communication system between the gut and the brain. The gut possesses its own nervous system, known as the enteric nervous system, commonly called the "second brain" due to the sheer number of nerve cells it possesses.

This connection works through:

- Nerves (especially the vagus nerve)
- Hormones
- Immune signals
- Gut microbes (which produce mood-influencing chemicals like serotonin)

This means that our mood, thoughts, and mental health can affect our digestion, and gut problems can also influence our emotions, leading to anxiety, depression, or brain fog.

Introduction to Prebiotics, Probiotics, and Postbiotics

The human gut is inhabited by a sophisticated collection of microorganisms that are essential in ensuring well-being. Among these, prebiotics, probiotics, and postbiotics are integral parts that ensure gut health and well-being. Knowing their definitions, roles, and examples will assist one in understanding the importance of these in health and nutrition.

Prebiotics

Definition: Prebiotics are undigestible food ingredients, usually fibers that stimulate the growth and activity of good gut bacteria. They are a source of food for probiotics, which helps to make them more effective.

Function: By promoting the growth of beneficial bacteria, prebiotics may enhance gut health, increase mineral absorption, and aid immune function. They also help produce short-chain fatty acids (SCFAs), which have numerous health benefits.

Examples:

- Inulin: Found in foods like chicory root, garlic, and onions, inulin is a soluble fiber that promotes the growth of beneficial bacteria.
- Fructooligosaccharides (FOS): Present in fruits and vegetables such as bananas and asparagus, FOS helps stimulate the growth of beneficial gut bacteria.
- Galactooligosaccharides (GOS): Found in dairy products and legumes, GOS supports the growth of *bifidobacteria,* which are important for gut health.

Probiotics

Definition: Probiotics are live bacteria and yeast that are beneficial when consumed in adequate amounts. They promote a healthy gut microbiome.

Function: Probiotics can enhance digestion, boost immune function, and possibly prevent or treat some gastrointestinal diseases. They can also synthesize beneficial products, such as vitamins that support well-being.

Examples:

- *Lactobacillus:* This genus includes various species found in yogurt and fermented foods, known for their ability to improve lactose digestion and support gut health.
- *Bifidobacterium:* Commonly found in the intestines, these bacteria are often included in probiotic supplements and are known for their role in maintaining gut health and preventing diarrhea.
- *Saccharomyces boulardii*: A beneficial yeast that can help prevent antibiotic-associated diarrhea and support gut health.

Postbiotics

Definition: Postbiotics are metabolic end products generated by probiotics through fermentation. They consist of several compounds like SCFAs, enzymes, and cell wall materials that can impart health benefits.

Function: Postbiotics may improve gut barrier function, regulate immune responses, and have anti-inflammatory effects. They may also affect the gut microbiome by imparting beneficial effects even after the probiotics have been eliminated.

Examples:

- Short-Chain Fatty Acids (SCFAs): Such as acetate, propionate, and butyrate, SCFAs are produced during the fermentation of dietary fibers by gut bacteria and play a crucial role in gut health and metabolism.
- Lactic Acid: Produced by lactic acid bacteria during fermentation, it can help lower the pH of the gut, creating an environment that inhibits harmful bacteria.
- Bacterial Cell Wall Components: These can stimulate immune responses and contribute to gut health.

Prebiotics, probiotics, and postbiotics are integral parts of gut health that all serve an individual purpose in maintaining the microbiome and health. Adding food containing these ingredients to your diet can create a healthy balance of gut microbiome and improve health results. Knowing their functions and foods that contain them can help people take active control over making the best dietary choices for improved digestion.

Chapter No 03

The Microbiome and Immunity

How microbes train and regulate our immune system

The human body is inhabited by trillions of microbes, particularly in the gut, where most immune cells also reside. The microbes are not only harmless passengers but actively and critically contribute to educating, training, and regulating the immune system during our lifetime.

Training the Immune System (Particularly in Early Life)

- At birth, the immune system is immature. As microbes start to populate the body via birth, breastfeeding, diet, and environment they "educate" the immune system to distinguish friendly microbes from injurious invaders.

- Beneficial microbes assist immune cells in learning tolerance, and thus the body won't attack harmless things like food proteins or good bacteria. Without adequate exposure to microbes in early life, the immune system can end up being over reactive and produce allergies, asthma, or autoimmune diseases.

Strengthening the Gut Barrier

- Microbes aid the integrity of the gut lining, which is a protective wall between the outside world and the body.

- The gut-friendly bacteria induce the release of mucus and tight junction proteins that do not allow toxins to enter the bloodstream. An intact barrier diminishes chronic inflammation and keeps the immune system from being over-activated.

Synthesizing Immune-Stimulating Compounds

Microbes synthesize vital compounds that influence immunity directly:

- Short-chain fatty acids such as butyrate, synthesized when microbes ferment dietary fiber, decrease inflammation, sustain immune cells, and heal gut tissue.

- They also stimulate regulatory T cells, which help the immune system stay balanced and avoid attacking the body's own cells.

Controlling Harmful Microbes

Good microbes compete with harmful ones by:

- Taking up space and nutrients so pathogens can't grow easily
- Producing antimicrobial compounds that kill or suppress harmful bacteria
- Signaling immune cells to respond when threats appear

This keeps the immune system alert but not overactive, ensuring a balanced response.

Communication with Immune Cells

Microbes transmit chemical messages to immune cells in the lining of the gut. They:

- Contribute to activating immune responses when necessary
- Encourage the building of immune tolerance to non-harmful substances
- Facilitate the growth of certain immune cells, such as B cells and T cells, which combat infection

Briefly, our microbes are silent teachers to our immune system. They teach it what to fight, what to tolerate, and how to remain in balance. Without them, the immune system would be unable to act properly, exposing the body to greater risk from disease, inflammation, and immune disorders.

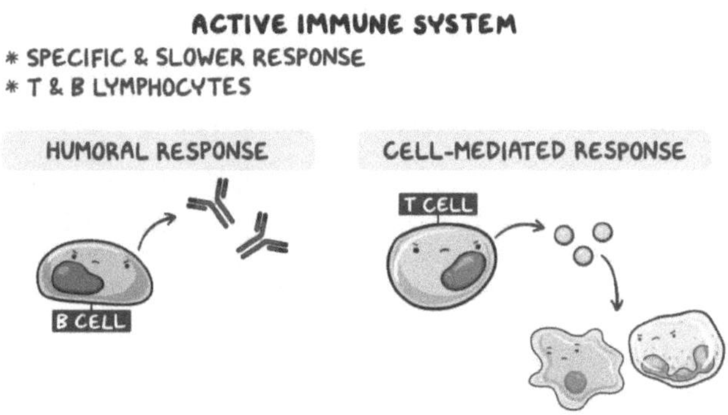

Figure 9: Shows immune response

Role in preventing infections by outcompeting harmful microbes:

Our body, particularly the gut, skin, mouth, and other surfaces, is inhabited by millions of beneficial microbes. These beneficial microbes, commonly referred to as commensals or probiotics, create a protective layer that guards us against harmful pathogens.
One of their most crucial defense mechanisms is "competitive exclusion", i.e., they outcompete for bad microbes and prevent them from growing and inflicting infections.

How Beneficial Microbes Outcompete Harmful Microbes?

Occupying Space

Beneficial microbes stick to the lining of the gut and other tissue. By doing so, they prevent bad bacteria from having room to adhere and proliferate. This is similar to booking all the seats at a playhouse so the misbehavers can't get a seat.

Using Up Nutrients

Beneficial microbes use up the available nutrients, including sugars and amino acids, leaving hardly any for pathogens to live on. This competition for nutrients starves the harmful microbes so that they cannot reproduce.

Producing Antimicrobial Substances

Certain beneficial bacteria release natural chemicals such as:

Bacteriocins proteins that kill or inhibit closely related pathogenic bacteria. Hydrogen peroxide particularly by vaginal bacteria such as *Lactobacillus*, which kills yeast and pathogens. Organic acids (such as lactic acid and acetic acid) – which decrease pH and make an environment in which pathogens are unable to survive.

Strengthening the Immune System

Beneficial microbes assist in training and priming immune cells, which are able to recognize and destroy invading pathogens with speed. A robust, microbe-aided immune system is better tuned and more potent in combating infections.

Gut Barrier Integrity

Good microbes maintain the gut lining, holding back harmful bacteria and their toxins from entering the bloodstream. This shields the body from inflammation and systemic infection.

Examples in Everyday Life:

• In the digestive system, *Lactobacillus* and *Bifidobacterium* control pathogenic bacteria such as *Salmonella* and *Clostridium difficile.*

• On the skin surface, beneficial bacteria such as *Staphylococcus epidermidis* maintain a balance and prevent overpopulation of deadly strains such as *Staphylococcus aureus.*

• In the vaginal tract, *Lactobacillus* prevents yeast and pathogens from developing into infections.

Helpful microbes serve as bodyguards in the natural order of things, preventing infection by filling space, competing for resources, and generating defensive chemicals. They ensure that a balance exists within our body and prevent illness with reduced risk of disease, which makes them a major component of our defense mechanism in nature.

Dysbiosis (disease state) and its influence on autoimmune illness, allergies, and inflammation:

Overview of Dysbiosis

Dysbiosis is a term applied when there is disturbance of the natural balance of microbes in the body particularly in the gut. It occurs when friendly microbes are reduced and pathogenic microbes increase, resulting in a loss of microbial diversity. This imbalance can adversely affect the immune system, bringing with it a variety of health issues.

Influence on Autoimmune Diseases

In autoimmune illnesses, the immune system attacks the body's own cells mistakenly. Dysbiosis can:

- Initiate defective immune signals, leading to the loss of the body's ability to differentiate between self and foreign.
- Permit leaky gut, wherein toxic substances flow through the gut lining and into the bloodstream, potentially triggering autoimmunity.
- Be associated with diseases such as type 1 diabetes, rheumatoid arthritis, multiple sclerosis, and lupus.

Influence on Allergies

A balanced gut microbiota educates the immune system to not react to harmless things such as pollen or food proteins. When this equilibrium is disrupted:

- The immune system might over respond to harmless stimuli, leading to allergy.
- Dysbiosis in early childhood has been linked to asthma, eczema, and food allergies.
- Lack of beneficial microbes, particularly in children, can predispose to allergy.

Impact on Inflammation

- Dysbiosis tends to result in chronic low-grade inflammation, which is involved in the majority of diseases. Pathogenic microbe production of SCFAs, typically anti-inflammatory, is decreased and worsens this condition. This inflammatory state is associated with obesity, cardiovascular disease, IBS, and depression.

- Having a healthy microbial balance is vital to keeping the immune system under control. Dysbiosis will disrupt immune tolerance, resulting in autoimmunity, allergies, and chronic inflammation. Gut health can be promoted by diet, probiotics, and lifestyle and restore balance to protect long-term health.

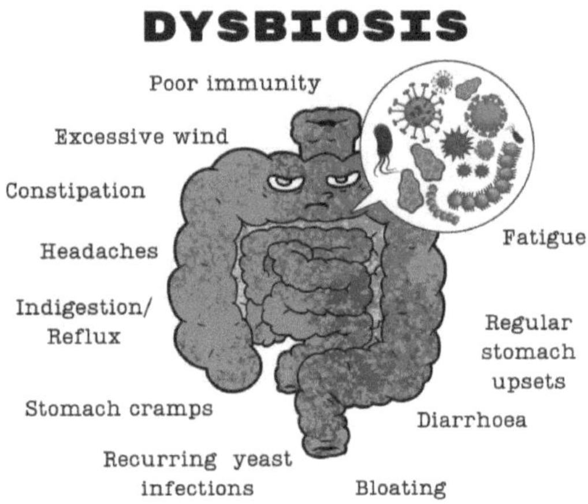

Figure 10: Shows dysbiosis (imbalance) and its impact

Microbiome's role in vaccine responses:

Microbiome, particularly gut microbiota, has an important function in determining how the body reacts to vaccines. Vaccines achieve this by triggering the immune system to develop a protective response, and it may be modulated by microbes residing within our body.

Boosting Immune Readiness

A well-functioning microbiome keeps the immune system on guard and in balance. Certain microbes in the gut can make antibody production more efficient, which is important for effective vaccine response.

Aiding Immune Cell Activation

Certain microbes assist in activating T cells and dendritic cells, the immune factors that are most responsible for recognizing vaccine elements and establishing long-term immunity.

Manufacturing Beneficial Molecules

Short-chain fatty acids (such as butyrate) and other compounds made by gut bacteria promote immune signaling and potentially increase vaccine efficacy.

Affecting Vaccine Efficacy

It has been revealed by science that individuals with varied and balanced microbiome respond more strongly and for a longer period of time to vaccines. In contrast, individuals with an unbalanced microbiome (dysbiosis) can present with a weaker immune response.

Age and Microbiome Relationship

In infants and the elderly, the composition of the microbiome influences the efficacy of vaccines. In infants, exposure to microbes early in life helps "school" the immune system. In older adults, a less rich diversity of the microbiome might lower response to vaccines.

The microbiome is a silent collaborator in ensuring vaccines become more effective. A healthy and balanced gut may result in enhanced immunity, not only to natural infections, but also to vaccines.

Chapter No 04

The Mind-Microbe Connection

The gut-brain axis: How gut microbes communicate with the brain?

Overview of the Gut-Brain Axis

The gut-brain axis (GBA) is a network that communicates between the gut and the brain. It maintains the digestive system functioning correctly, affects brain functions like mood, cognition, and mental state, and regulates immune responses. The most important members of the system are:

- **The Enteric Nervous System (ENS):** A dense network of neurons lining the gut, sometimes called the "second brain."

- **The Vagus Nerve:** The major nerve connecting the gut directly to the brain.

- **Gut Microbiota:** Trillions of microbes (bacteria, viruses, fungi) in the GI tract that play a critical role in maintaining homeostasis.

- **Immune Signaling and Hormones:** Cytokines and gut-derived hormones are also involved in this communication.

How Gut Microbes Communicate with the Brain

- **Via the Vagus Nerve**

The vagus nerve is a direct communication pathway between the gut and brain. The gut bacteria metabolize to form short-chain fatty acids (SCFAs), which can affect vagal nerve activity and, in turn, brain functions such as mood control and stress response.

- **Through Metabolites and Neurotransmitter Production**

Gut microbes can synthesize neurotransmitters like serotonin, dopamine, gamma-aminobutyric acid (GABA), and acetylcholine. For example, almost 90% of the body's serotonin is synthesized in the gut.

Microbial metabolites such as SCFAs (e.g., butyrate, propionate) are delivered to the brain through the bloodstream, where they influence brain function and inflammation status.

- **Immune System Modulation**

Gut microbes modulate immune cells and inflammatory cytokine production. Dysbiosis in the gut microbiota can generate neuroinflammation, which might be involved in neurodegenerative disease such as Alzheimer's or Parkinson's.

- **Endocrine Communication (Gut Hormones)**

The gut also secretes hormones, like ghrelin ("hunger hormone") and leptin ("satiety hormone") that affect brain areas responsible for appetite, mood, and cognitive processes.

- **Intestinal Barrier and Blood-Brain Barrier (BBB) Integrity**

Gut health affects intestinal lining permeability (toxic substances leak into the bloodstream when damaged) and the blood-brain barrier, which shields the brain. Both barriers are controlled by gut microbiota

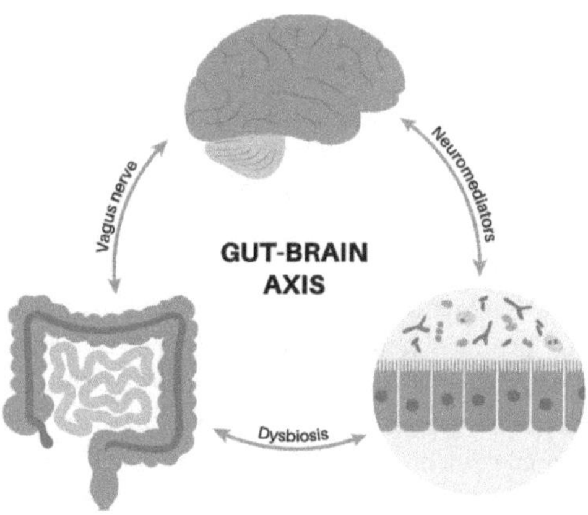

Figure 11: Shows how gut microbes communicate with the brain

Impact of Gut Microbes on Brain Functions

- **Mood and Mental Health**: The gut microbiota has also been implicated in conditions of anxiety, depression, and stress. For instance, probiotics (good bacteria) have shown to enhance mood and alleviate anxiety symptoms
- **Cognition and Behavior:** Research indicates that gut bacteria may affect memory and learning capabilities. Animal experiments, for instance, have established changed behaviors and cognitive impairment in germ-free mice (mice reared without any gut microbes).
- **Neurodegenerative and Neurological Disorders:** Dysbiosis has been associated with conditions such as autism spectrum disorder (ASD), multiple sclerosis (MS), and Alzheimer's disease. A certain microbiota profile could be involved in disease or symptom progression.

The gut-brain axis is a dynamic, complex network that highlights the interrelatedness of physical and mental well-being. Appreciation of how gut microbes talk to the brain presents opportunities for novel treatments aimed at the microbiota, including probiotics, prebiotics, dietary therapies, and even psychobiotics microbial strains with possible therapeutic benefits for mental health.

Role in mood, anxiety, depression, and even behavior

- **Gut Microbiota and Mood Regulation:**

The gut microbiota, which consists of trillions of microbes, affects the synthesis of neurotransmitters, hormones, and signaling molecules that are crucial for mood control. The main mechanisms are:

- **Production of Neurotransmitters:**

Microbes in the gut produce neurotransmitters such as serotonin (responsible for happiness), dopamine (pleasure and reward), and GABA (relaxation and stress relief). A considerable amount of serotonin up to 90% is synthesized in the gut, underlining its pivotal role in emotional control.

- **Short-Chain Fatty Acids (SCFAs):**

SCFAs such as butyrate and acetate, microbial fermentation by-products, affect brain processes by regulating inflammation, energy balance, and neurotransmitter release. SCFAs have been found to trigger the release of mood-improving molecules in the brain.

- **Vagus Nerve Connection:**

The vagus nerve is a direct communication pathway between the gut and brain. Messages from gut microbes shape brain function, affecting mood and emotions.

Gut-Brain Axis in Anxiety

The gut microbiota is significantly associated with anxiety disorders through mechanisms like:

- **Stress Response Regulation:** Gut microbes influence the hypothalamic-pituitary-adrenal (HPA) axis, which regulates the stress response. A healthy microbiota can prevent over activation of the HPA axis and decrease stress-induced anxiety.

- **Probiotics and Anxiety:** Research has demonstrated that some probiotics, popularly referred to as "psychobiotics," have been found to reduce anxiety. For example, *Lactobacillus* and *Bifidobacterium* strains are related to decreased cortisol levels and enhanced stress resilience.

The Role of Gut Microbiota in Depression

Dysbiosis (gut microbiota imbalance) has been linked with the development and severity of depression. This relationship stems from:

• **Chronic Inflammation:**

Dysbiosis may contribute to systemic inflammation, which in turn is associated with depressive symptoms. High levels of inflammatory cytokines, including IL-6 and TNF-α, are also often present in patients with depression.

•**Neuroplasticity:**

The gut microbiota controls brain-derived neurotrophic factor (BDNF) levels, a protein necessary for brain wellness and neuroplasticity. Low levels of BDNF correlate with depression.

•**Gut Permeability (Leaky Gut):**

Dysbiosis can interfere with the gut barrier, permitting toxic substances (e.g., lipopolysaccharides) into the bloodstream and inducing neuro inflammation and producing depressive symptomatology.

Influence on Behavior

Gut microbiota not only influences mood but behavior as well. Important observations are:

•**Cognitive Function:** Gut microbes modulate learning, memory, and decision-making by influencing neural circuits and neurotransmitter systems.

•**Social Behavior:** Experimental studies in animal models have demonstrated that modulating the gut microbiota can induce changes in social behaviors, such as sociability and aggression.

•**Microbiota-Gut-Brain Axis in Autism:** Gut dysbiosis is suspected to be involved in behavioral symptoms of autism spectrum disorders, including repetitive behaviors and impaired communication. Probiotics and dietary interventions are being investigated to enhance symptoms.

Therapeutic Potential

The gut-brain axis has therapeutic potential with the following:

- **Probiotics and Prebiotics:**

Probiotics (live healthy bacteria) can enhance mood and decrease anxiety and depression symptoms. For instance, *Lactobacillus rhamnosus* has been shown to have anti-anxiety effects in research. Prebiotics (undigestable fibers that feed gut bacteria) enhance the health of good bacteria and encourage SCFA production.

- **Dietary Interventions:**

Fiber-rich diets, fermented foods, and omega-3 fatty acid support a balanced gut microbiota and enhance mental health outcomes.

- **Fecal Microbiota Transplantation (FMT):**

Rising as a possible cure for extreme depression and anxiety, FMT is a process of transferring gut bacteria from a donor who is healthy in order to rebuild a balanced microbiota.

The gut-brain axis is transforming our knowledge of mental health. As we clarify the workings of gut microbes in mood, anxiety, depression, and behavior, we are discovering novel treatments for psychological disorders. Treating the gut microbiota with probiotics, diet, and other treatments is an exciting frontier in mental health treatment.

Link with neurological condition like Autism and Alzheimer's

What is Autism? Autism Spectrum Disorder (ASD)

Autism is a developmental disorder impacting communication, behavior, and social interaction. It is referred to as a "spectrum disorder" due to the variability in symptoms and their severity among different individuals.

Symptoms of Autism:

- Social Challenges: Inability to comprehend social cues and form relationships. Refusal to maintain eye contact and preference for being alone.
- Communication Difficulties: Delay in or absence of speech development. Inability to express emotions or needs, resorting to repetitive phrases.
- Behavioral Patterns: Stereotypic movements (hand-flapping, rocking). Strong interest in or preoccupation with specific topics or rituals. Struggle to accommodate shifts in setting or schedule.
- Sensory Sensitivities: Overreaction or under reaction to sensory stimulation, including sounds, lights, or textures.

Gut-Brain Axis in Autism Spectrum Disorder (ASD)

- **Gut Dysbiosis and Behavioral Symptoms:**

Research has identified that children with autism tend to have abnormal gut microbiota composition versus neurotypical people. An imbalance of gut bacteria could be a cause of gastrointestinal symptoms that are common in people with ASD and increase the difficulty of behavioral challenges. Dysbiosis can interfere with neurotransmitter synthesis and enhance inflammatory signaling, affecting brain function and behavior.

- **Immune Dysregulation:**

The gut microbiota influences the immune system, which tends to be deregulated in autism. Low-grade chronic inflammation from the gut may influence neural development and connectivity.

- **Metabolites and Neurodevelopment:**

Some gut microbes produce metabolites that are essential for brain development. For instance, short-chain fatty acids (SCFAs) promote neuroplasticity and immune health. An absence of beneficial SCFA-producing bacteria may impede neurodevelopment.

- **Therapeutic Approaches:**

Interventions aimed at the gut microbiota are also under investigation, including probiotics, prebiotics, and diet modification. Fecal microbiota transplantation (FMT) is

also being investigated as a possible treatment for enhancing gut function and reducing symptoms in autism.

What is Alzheimer's? Alzheimer's disease

Alzheimer's disease is a degenerative neurological condition and the leading cause of dementia. It causes declining cognitive function, memory, and impaired capability for everyday activities.

Symptoms of Alzheimer's:

- Early Symptoms: Trouble remembering recent events or conversations. Forgetting appointments and misplacing items.
- Progressive Cognitive Decline: Difficulty solving problems or planning tasks. Trouble recognizing familiar places or faces. Disorientation in space and time.
- Behavioral Changes: Mood changes, depression, or anxiety. Personality changes and social withdrawal. Restlessness and agitation in later stages.
- Severe Stages: Loss of communication ability. Needing caregivers for everyday activities. Trouble swallowing or moving.

Gut-Brain Axis in Alzheimer's disease

- **Chronic Inflammation and Neurodegeneration:**

Gut microbiota dysbiosis is associated with chronic systemic inflammation, which is a feature of Alzheimer's disease. Increased production of inflammatory cytokines from the gut can augment neurodegeneration and cognitive impairment.

- **Gut Microbiota and Amyloid-Beta Accumulation:**

Some bacteria generate metabolites or toxins that can favor the formation of amyloid-beta plaques in the brain. These plaques are typical of Alzheimer's pathology and correlate with cognitive impairment.

- **Gut Barrier and Blood-Brain Barrier Integrity:**

Dysbiosis disrupts the lining of the gut, resulting in "leaky gut." It permits noxious substances to be absorbed into the bloodstream and activate neuroinflammation. Just like dysregulation of the blood-brain barrier may aggravate Alzheimer's pathology.

- **Impact on Cognitive Functions:**

Modification of gut microbiota composition alters brain energy metabolism and neurotransmitter synthesis, producing cognitive impairment, including memory impairment and impaired learning.

- **Therapeutic Strategies:**

Probiotics and nutritional therapy for restoring gut microbial balance are also being researched as treatment options. Certain dietary factors, such as polyphenols in fruits and vegetables, have been shown to limit inflammation and promote brain health. FMT and other microbiota-directed therapies are being investigated for their potential to enhance gut-brain communication and decelerate disease progression.

The gut-brain axis is a critical factor in understanding the underlying mechanisms of autism spectrum disorder and Alzheimer's disease. By targeting the gut microbiota through innovative therapies like probiotics, prebiotics, dietary interventions, and FMT, researchers hope to improve symptoms and slow the progression of these neurological conditions. This field of study is transforming our approach to brain health and offers promising avenues for treatment.

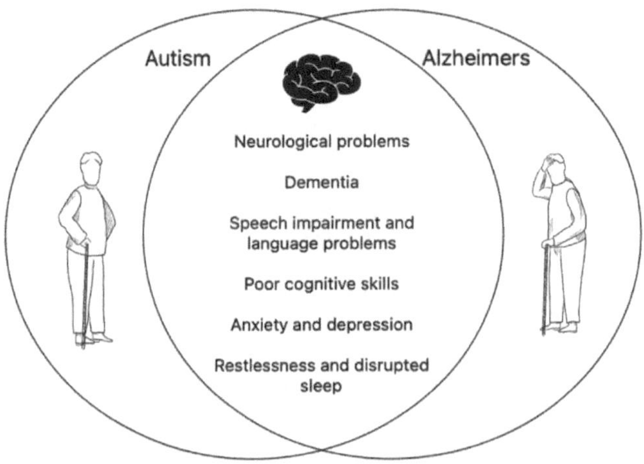

Figure 12: Shows the symptoms of both diseases

How diet and lifestyle affect this connection?

Impact of Diet on Gut-Brain Axis

- **Microbiota Composition:**

Diet regulates the diversity and content of gut microbiota, which are central to the gut-brain axis. Fiber-rich, fruit, vegetable, and fermented food diets enhance the proliferation of beneficial microbes such as Lactobacillus and *Bifidobacterium,* which are beneficial for mental and neurological well-being. On the other hand, high-processed food, sugar, and unhealthy fat diets may cause dysbiosis (imbalance of microbes), leading to inflammation and disrupted communication between the gut and brain.

- **Production of Major Metabolites:**

Short-chain fatty acids (SCFAs): Diets rich in fiber stimulate gut microbiota to make SCFAs (e.g., butyrate), lowering inflammation and favoring brain well-being.

- **Neurotransmitters:**

Diet determines microbial creation of neurotransmitters like serotonin and dopamine. Foods high in tryptophan (e.g., eggs, nuts), for instance, provide substrates that facilitate serotonin creation.

- **Influence on Intestinal Barrier:**

Healthy eating supports the integrity of the gut lining, avoiding "leaky gut," where dangerous substances can enter the blood and influence the brain.

- **Anti-Inflammatory Foods:**

Omega-3 fatty acids (in fish, walnuts, flaxseeds) contain anti-inflammatory nutrients that are beneficial for both gut and brain health. They suppress brain plasticity and protect against neuroinflammation.

- **Probiotics and Fermented Foods:**

Dietary foods such as yogurt, kimchi, and kefir harbor live healthy bacteria that directly increase the microbial diversity of the gut and enhance gut-brain communication.

Influence of Lifestyle on Gut-Brain Axis

• Stress Management:

Long-term stress dysregulates the hypothalamic-pituitary-adrenal (HPA) axis and adversely affects the gut microbiota, inducing inflammation and increased gut permeability.

Stress-reducing interventions such as meditation, yoga, and slow breathing enhance microbiota health and gut-brain communication.

• Sleep Quality:

Poor sleep interferes with gut microbiota diversity and gut-brain communication. Proper sleep restores microbial balance and facilitates the production of neurotransmitters that control mood and cognitive processes.

•Physical Activity:

Exercise enhances gut microbiota diversity and boosts SCFA production, which improves brain health and alleviates symptoms of anxiety and depression.

•Avoiding Harmful Substances:

Heavy drinking and smoking have adverse effects on gut health by decreasing microbial diversity and inducing inflammation, which can impair brain function.

•Hydration:
Hydration maintains gut barrier function and enhances the movement of nutrients and metabolites vital to brain health.

Specific Connections to Neurological and Mental Health Conditions

- **Autism Spectrum Disorder (ASD):**

Fiber and probiotic diets can enhance gut health and help reduce some ASD symptoms, like gastrointestinal distress and mood disturbances. Anti-inflammatory, nutrient-rich diets promote neurodevelopment and behavior.

- **Alzheimer's disease:**

A Mediterranean diet high in omega-3 fatty acids, antioxidants, and fiber is linked with lowered risk of Alzheimer's by ensuring good gut and brain health. Lifestyle modifications such as exercise and stress reduction are instrumental in lowering inflammation and cognitive impairment.

Nutrition and lifestyle are pivotal to the maintenance of a healthy gut-brain axis. Through enhanced microbial diversity and decreased inflammation through nutrient-rich diets, appropriate exercise, stress reduction, and good sleep habits, mental and neurological well-being is enhanced. Such changes provide promising prevention and management strategies for conditions such as anxiety, depression, autism, and Alzheimer's disease.

Chapter No 05

Microbiome Disruption-Causes and Consequences

Factors that disturb the human microbiome:

The human microbiome is a complex group of microorganisms living mainly in the gut, and it is responsible for digestion, immunity, and general health. However, some external and internal factors may interfere with the balance, and as a result, dysbiosis occurs. Some of the most significant contributors to microbiome disruption are listed below:

- Antibiotics
- Diet,
- Stress,
- C-section birth,
- Sanitization

• Antibiotics

Antibiotics are formulated to kill bad bacteria in the course of infections, but they tend not to differentiate between harmful and beneficial microbes. Consequently, frequent or long-term antibiotic use can significantly deplete microbial diversity, eradicating useful bacteria and paving the way for resistant or opportunistic organisms to overgrow. Such imbalance may elevate the risk of infections, gastrointestinal problems, and immune-related diseases.

• Diet

Diet has a profound impact on the composition and diversity of the microbiome. Diets high in processed foods, sugar, and low in fiber tend to support the growth of harmful bacteria, while suppressing beneficial ones. In contrast, a diverse, plant-rich diet provides prebiotics and nutrients that promote the growth of helpful microbes. Sudden dietary changes can also cause temporary imbalances in microbial populations.

• Stress

Psychological stress affects not only the mind but also changes the gut environment. Chronic stress can alter gut motility, immune function, and hormone levels, which all play a role in affecting the microbial community. Stress-induced changes to the gut lining can make it more permeable ("leaky gut"), permitting harmful substances to access the blood and triggering inflammation that further destabilizes microbial balance.

- **Cesarean Section (C-Section) Birth**

Infants delivered through cesarean section lack exposure to their mother's vaginal and intestinal microbiota during delivery. Instead, they are colonized initially by hospital environment and cutaneous microbes, which can hinder the establishment of a stable and diverse microbiome. Such initial microbial imbalance has been associated with higher risks of allergies, asthma, and metabolic diseases in the future.

- **Excessive Sanitization**

Although hygiene is important for disease prevention, excessive use of disinfectants and antibacterial products can restrict exposure to natural microbes that are important for immune system training. Especially during early childhood, exposure to a range of environmental microbes is important in developing a healthy and tolerant immune response. Over-sanitization can thus interfere with microbial diversity and lead to autoimmune and allergic diseases.

What are the long term effects of dysbiosis?

Dysbiosis is a disruption of the balance of the body's microbial populations, particularly in the gut. This disruption of microbial diversity and function can have significant implications for physical and mental well-being. When dysbiosis is chronic or unresolved, it can play a role in the development or exacerbation of many long-term health disorders.

Chronic Inflammatory Conditions

An imbalanced microbiome is able to induce and perpetuate low-grade inflammation throughout the body. This ongoing inflammation is a causative component in many chronic diseases, including:

- Inflammatory bowel disease (IBD) like Crohn's disease and ulcerative colitis.
- Irritable bowel syndrome (IBS), frequently being related to disrupted gut flora.
- Rheumatoid arthritis and other autoimmune diseases, as microbial imbalances can interfere with immune tolerance.
-

Metabolic Disorders

Dysbiosis can disrupt the way the body absorbs nutrients and stores energy. Overgrowth of some gut bacteria can pull extra energy from food, leading to fat storage and insulin resistance. This imbalance has been associated with:

- Obesity
- Type 2 diabetes
- Non-alcoholic fatty liver disease (NAFLD)

These conditions can get worse over time if the gut environment is not healthy.

Mental Health Impacts

The brain and gut are intimately linked via the gut-brain axis. Dysbiosis can influence the manufacture of neurotransmitters such as serotonin, affecting behavior and mood. Dysbiosis over time has been linked with:

- Depression
- Anxiety
- Cognitive decline and possibly neurodegenerative conditions like Alzheimer's disease

Weakened Immunity

The microbiome helps to modulate immune reactions. An impaired gut environment may dampen the ability of the immune system to discern between injurious invaders and one's own body tissues. Such impairment may progressively result in:

- Frequent infections
- Higher levels of allergies
- Autoimmune diseases like lupus or multiple sclerosis

Higher Risk of Cancer

Chronic inflammation and the generation of toxic bacterial metabolites during dysbiosis have the potential to cause DNA damage, interfere with cellular regulation, and induce tumor growth. Long-term microbial imbalance has been associated with:

- Colorectal cancer
- Stomach cancer
- Breast and liver cancers possibly, although more studies are underway

Skin and Respiratory Disorders

The consequences of dysbiosis may also be felt outside the gut in other body systems:

- Skin conditions such as eczema, acne, and psoriasis have been linked to changes in gut bacteria.
- Asthma and allergic rhinitis can also be affected by early-life microbial imbalance

Dysbiosis is not just a gut problem it has systemic effects that will affect almost all organ systems. Restoring balance to the microbes through diet, probiotic supplementation, and lifestyle modifications is vital to prevention of these chronic diseases and maintaining overall health.

Case Study: The unsuspected effect of Gut imbalance

Name: Rameen Ali
Age: 32
Occupation: Software Engineer
Location: Lahore, Pakistan

Background

Rameen was a successful, vibrant woman with a hectic lifestyle. As with many professionals, fast food, soft drinks, and late-night eating were often her diet. She used to take antibiotics on a regular basis for minor infections, especially throat and urinary tract infections. For a long time, she began to develop frequent digestive disturbances bloating, irregular bowels, and stomach pain. At first, she wrote them off as stress or merely bad eating days.

Symptoms and Medical History

In her early 30s, Rameen noticed changes in her overall well-being:

- Persistent fatigue following well-rested sleep
- Mood changes, anxiety, and sporadic panic attacks
- Weight gain though her eating habits hadn't varied much
- Skin troubles such as acne in adults and dry spots
- Recurrent colds and intolerance to foods
-

She had consulted several physicians who addressed individual symptoms. There were some doctors who prescribed anti-depressants, while some gave her creams or probiotics with no underlying diagnosis.

Breakthrough

Ultimately, she went to a holistic doctor who recommended an exhaustive gut health screening. Results confirmed severe dysbiosis her gut microbiome was grossly unbalanced with prevalence of inflammatory bacteria and the absence of probiotics such as *Lactobacillus* and *Bifidobacterium.* Her previous consistent use of antibiotics, processed food, and stress were seen to be the principal contributors.

Treatment and Recovery

Rameen was prescribed a 6-month gut retraining program consisting of:

- A fiber-rich, whole-food diet with fermented foods and prebiotics
- Specific probiotic supplements to repopulate beneficial bacteria
- Stress reduction through gentle exercise and mindfulness
- Removal of inflammatory triggers such as gluten and dairy for a short time

Gradually, she saw remarkable changes not only in digestion, but also in her mental acuity, energy, and skin. Her anxiety attacks diminished, and she no longer had to depend on regular medicines. She became a vocal promoter of gut health consciousness among her friends.

Takeaway

Rameen's experience illustrates a widespread but underappreciated problem: how imbalance in the gut can express itself in ways that impact the entire body. What started as mild digestive distress became a systemic problem, but with proper knowledge and attention, she was able to reverse much of the damage.

Figure13: Case Study

Emerging Issue: Decline in microbial diversity in contemporary lifestyle

Over the past few years, scientists and medical professionals have sounded the alarm about a quiet but profound change in human health: the loss of microbial diversity as a result of contemporary lifestyle. Our bodies, and especially our intestines, are home to trillions of microorganisms that are essential for digestion, immunity, brain function, and disease prevention. But the diversity and richness of those microbes are dwindling, and the effects are becoming more and more apparent.

What is Microbial Diversity?

Microbial diversity is used here to denote variety of microbial species found in the human body. A healthy gut, for instance, hosts hundreds of various bacterial strains, each carrying out distinct and critical functions. Increased diversity tends to imply increased resistance to diseases, more equilibrated immunity, and stable metabolic function.

How the Modern Lifestyle is contributing to Microbial Loss?

1. Highly Processed Diets

Contemporary diets tend to be low in fiber and high in sugar, fat, and additives. Such foods feed only a small subset of microbes but starve the beneficial microbes that live on plant fibers and natural nutrients.

2. Excessive Use of Antibiotics

Although antibiotics are crucial in the treatment of infections, their unnecessary and repeated use has greatly damaged the microbiome. Antibiotics eliminate both beneficial and deleterious bacteria, causing long-term imbalances and lower microbial richness.

3. Decreased Exposure to Natural Environments

City living, minimal soil contact, and less outdoor time have significantly lowered our exposure to environmental microbes. Kids today grow up in cleaned-up homes, play less outside, and get fewer chances to establish a strong microbial foundation.

4. Over-sanitization and Hygiene Products

Overuse of antibacterial soaps, disinfectants, and cleansers, although they are infection-protective, also destroy harmless and helpful microbes on our skin, hands, and surroundings. This can be a contributing factor to increased allergies and autoimmune diseases.

5. Cesarean Deliveries and Formula Feeding

Infants delivered by C-section and those not breastfed lack essential microbes from the mother's body and milk. These initial exposures are important for developing a robust and diverse microbiome.

6. Sedentary Lifestyle and Chronic Stress

Chronic stress and lack of activity can change gut motility, hormone levels, and immune function, all of which have negative effects on microbial balance and decrease diversity over time.

Why This Loss is Alarming

Loss of microbial diversity compromises the body's capacity to adapt and defend against disease. It has been associated with an increase in:

- Autoimmune diseases
- Allergies and asthma
- Obesity and diabetes
- Depression and anxiety
- Gastrointestinal disorders such as IBS and IBD

Adding to the alarm is that some microbial species, once lost, cannot be regained without assistance.

A Call for Microbiome-Friendly Living

To turn this around, there need to be conscious shifts in lifestyle, such as:

- Consuming a wide range of whole, plant-based foods
- Spending time outdoors and gardening
- Limiting unnecessary antibiotics
- Employing natural, non-toxic household cleaners
- Supporting breastfeeding and natural childbirth when possible
- Managing stress and maintaining physical activity

The loss of microbial diversity is a recent health threat with entrenched causes and far-reaching implications. As the word spreads, adopting microbiome-friendly lifestyles might become an integral component of future public health policy. Taking care of our microbes isn't solely about digestive health it's about maintaining a part of what makes us human.

The microbiome plays a key role in regulating immune responses. A disrupted gut environment can impair the immune system's ability to distinguish between harmful invaders and the body's own tissues. Over time, this can lead to:

- Frequent infections
- Increased allergies
- Autoimmune diseases such as lupus or multiple sclerosis

Increased Risk of Cancer

Chronic inflammation and the production of harmful bacterial metabolites during dysbiosis can damage DNA, disrupt cellular regulation, and promote tumor growth. Long-term microbial imbalance has been linked to:

- Colorectal cancer
- Stomach cancer
- Possibly breast and liver cancers, though more research is ongoing

Skin and Respiratory Conditions

The effects of dysbiosis can extend beyond the gut to other body systems:

- Skin disorders like eczema, acne, and psoriasis have been associated with altered gut bacteria.
- Asthma and allergic rhinitis may also be influenced by early-life microbial imbalance

Long-term dysbiosis is not merely a gut issue it has systemic consequences that can impact nearly every organ system. Restoring microbial balance through a healthy diet, probiotic use, and lifestyle changes is essential to reducing the risk of these chronic conditions and supporting overall health.

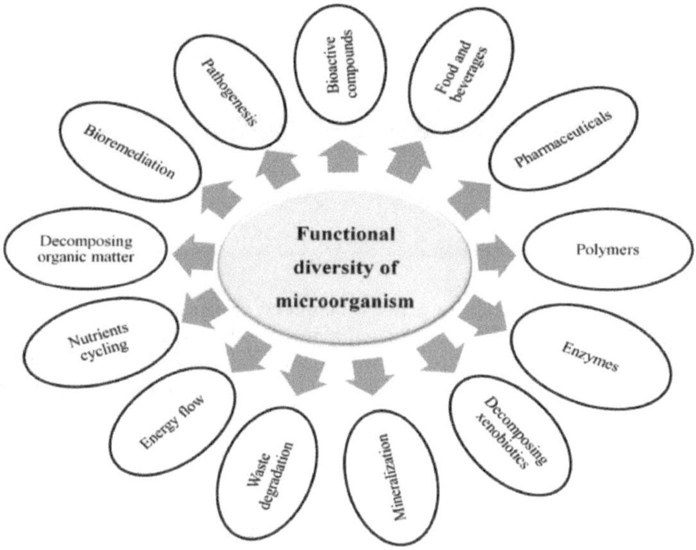

Figure 14: Shows microbial diversity and their functions

Chapter No 06

Protecting and Nurturing your Microbiome

Simple ways to maintain a healthy microbiome:

Maintaining a healthy microbiome can be achieved through several simple practices. Incorporate a diverse diet rich in fruits, vegetables, whole grains, and fermented foods like yogurt and kimchi, as these promote beneficial bacteria. Additionally, staying hydrated, managing stress, and avoiding unnecessary antibiotics can further support microbiome health.

Eat a Diverse Range of Foods

- **Fruits and Vegetables:** Aim for a variety of colors and types to ensure a wide range of nutrients.
- **Whole Grains:** Oats, quinoa, and brown rice offer fiber that acts as a food source for beneficial gut bacteria.
- **Fermented Foods:** Add yogurt, kefir, sauerkraut, and kimchi, which have live probiotics to boost gut health.

Stay Hydrated

- **Water Intake:** Drink enough water during the day to aid digestion and nutrient absorption.
- **Hydration and Fiber:** Proper hydration ensures fiber can perform its functions effectively, with regular bowel movements and a healthy gut.

Manage Stress

- **Mindfulness Practices:** Engage in activities like yoga, meditation, or deep-breathing exercises to reduce stress levels.
- **Physical Activity:** Regular exercise can positively influence gut health by promoting the growth of beneficial bacteria.

Limit Antibiotic Use

- **Use Only When Necessary:** Avoid taking antibiotics unless prescribed by a healthcare professional, as they can disrupt the balance of gut bacteria.

- **Probiotics during Antibiotic Treatment:** If antibiotics are necessary, consider taking probiotics to help restore gut flora afterward.

Avoid Processed Foods

- **Restrict Sugar and Artificial Additives:** Consuming excessive sugars and artificial components can harm gut health.
- **Whole Foods:** Give preference to whole, unprocessed foods to feed your microbiome well.

Get Enough Sleep

- **Sleep Hygiene:** Get 7-9 hours of sound sleep every night to encourage overall health, including gut health.
- **Standard Sleep Schedule:** Stick to a fixed sleep schedule to aid in synchronizing your body's internal clock.

Take Prebiotics into Account

- **Prebiotic Foods:** These foods include garlic, onions, bananas, and asparagus, which provide nutrients for good gut bacteria.
- **Supplementation:** If required, take prebiotic supplements after consulting with a healthcare professional.

Figure 15: Shows how to maintain healthy gut

Diet tips: Fiber, Fermented Foods, Diverse plant intake

Here's a comprehensive look at how fiber, fermented foods, and a diverse plant intake help make up a healthy diet and microbiome.

1. Fiber

Importance of Fiber: Fiber is a carbohydrate that cannot be digested by the body. It is essential for maintaining healthy guts by promoting normal bowel movements and nourishing healthy gut bacteria.

Types of Fiber:

- **Soluble Fiber:** Present in oats, beans, lentils, apples, and citrus fruits. Soluble fiber dissolves in water to produce a gel-like consistency, helping lower cholesterol levels and control blood sugar levels.
- **Insoluble Fiber:** In whole grains, nuts, seeds, and the fruit and vegetable skins. Increases bulk in the stool and accelerates the passage of food through the digestive tract.

Benefits of Fiber:

- **Supports Healthy Digestion:** The regular consumption of fiber prevents constipation and promotes digestive well-being.
- **Maintains Gut Microbiome:** Fiber serves as a prebiotic, giving nutrition to friendly bacteria in the gut, which can boost their numbers and activity.
- **Lowers Chronic Disease Risk:** High-fiber diets have a lower risk of heart disease, diabetes, and cancer.

Tips for Increasing Fiber Intake:

- Slowly introduce more fiber in the diet to prevent digestive distress.
- Mix various high-fiber foods with each meal, including fruits, vegetables, whole grains, and legumes.

2. Fermented Foods

What Are Fermented Foods? Fermented foods are foods that have been subjected to a process of lacto-fermentation, where natural bacteria break down the sugar and starch in the food, producing beneficial probiotics.

Examples of Fermented Foods:

- **Dairy:** Yogurt, kefir, and cheese.
- **Vegetables:** Sauerkraut, kimchi, and pickles.
- **Beverages:** Kombucha and fermented teas.

Benefits of Fermented Foods:

- **Probiotic Content:** Fermented foods contain high levels of probiotics, which are live microorganisms that can confer health benefits when taken in adequate amounts.
- **Enhanced Nutrient Absorption:** Fermentation can increase the bioavailability of certain nutrients, making them easier for the body to absorb.
- **Gut Health:** Regular consumption of fermented foods can help balance the gut microbiome, potentially improving digestion and immune function.

Tips for Incorporating Fermented Foods:

- Start with small portions to allow your gut to adjust.
- Experiment with different types of fermented foods to find what you enjoy.

3. Diverse Plant Intake

Importance of Plant Diversity:
A diverse diet with many different plant foods can greatly promote gut health and overall health. Various plants offer distinct nutrients and phytochemicals that help maintain a healthy microbiome.

Benefits of Diverse Plant Intake:

- **Nutrient Variety:** Various fruits, vegetables, grains, and legumes provide a variety of vitamins, minerals, and antioxidants necessary for health.

- **Microbiome Diversity:** Increased diversity in food can encourage increased diversity in gut microbiome, which is related to improved health and disease resilience.
- **Reduced Inflammation:** Plant foods are rich in anti-inflammatory molecules that may mitigate inflammation within the body.

Increase Plant Diversity:

Try to consume a variety of colors on the plate since diverse colors tend to represent different nutrients.

- Attempt new fruits and vegetables on a weekly basis to diversify your palate and nutrient profile.
- Eat a range of legumes, nuts, seeds, and whole grains.
- Adding fiber, fermented foods, and a variety of plant foods to your diet can have a tremendous impact on your gut health and overall health. With an emphasis on these aspects, you can nourish a healthy microbiome, enhance digestion, and minimize the risk of chronic diseases.
- Always make gradual changes and pay attention to your body while making adjustments to your eating habits.

Use of antibiotics and hygiene balance in a responsible manner

The human microbiome specifically the microbial community living in the gut—is critical to staying healthy. It aids digestion, regulates the immune system, defends against disease-causing pathogens, and even affects mood and brain function. But two contemporary tendencies excessive use of antibiotics and too much cleanliness—are serious dangers to the subtle balance of this internal ecosystem.

In order to protect and care for your microbiome, a wise and balanced strategy is important.

1. Proper Use of Antibiotics

Antibiotics are lifesaving drugs, but their overuse and misuse can deeply destroy the microbiome. Such drugs cannot differentiate between bad bacteria and good bacteria eradicating both and frequently resulting in a less diverse, weaker microbial population.

How to Properly Use Antibiotics:

- **Use them only when prescribed:** Do not take antibiotics for viral infections such as the common cold or flu they will not work and will be harmful.
- **Follow the full course:** Stopping antibiotics early can lead to resistance and incomplete recovery, giving harmful bacteria a chance to rebound.
- **Never self-medicate:** Taking leftover antibiotics or sharing them with others increases the risk of inappropriate use.
- **Ask about alternatives:** In mild infections, your doctor may recommend natural recovery or probiotics alongside treatment.

Why It Matters:

Excessive or unneeded antibiotic consumption has also been associated with long-term dysbiosis, compromised immunity, greater allergies, and even metabolic and mood illness.

2. Hygiene Balance: Clean, Not Sterile

While it's important that we practice good hygiene to avert infectious sickness, excessive sanitation will destroy good microbes we want to remain in health. New environments particularly city homes tend to be so germ-free that we reduce exposure to normal microbial amounts, especially children.

Healthy Hygiene Practices:

Wash hands with plain soap and water: Don't use antibacterial soaps unless absolutely necessary (e.g., in hospitals or during outbreaks).

Let children get dirty: Getting dirty in soil, with pets, and outdoors helps build a robust immune system.

Use disinfectants judiciously: Disinfect only surfaces in high-risk areas such as kitchens or bathrooms. Excessive use in low-risk areas can cause more harm than benefit.

Select natural cleaning products: They tend to be less damaging to the environment and your microbial environment.

Finding the Balance: Clean Enough, Not Over cleaned

The intention is not to eschew hygiene but to apply it consciously. A clean environment shields from disease, but a sterile one could deprive the microbiome of the variety it requires to flourish. In the same vein, antibiotics are lifesavers but should be respected and used judiciously lest they cause collateral damage to our internal microbial friends.

The first step in preserving and cultivating your microbiome is awareness. Paying attention to how we utilize antibiotics and how we sanitize our environment can help us foster beneficial microbes instead of inadvertently eradicating them. By doing this, we are investing in improved immunity, robust mental and metabolic well-being, and overall resilience for ourselves and generations to come.

Role of probiotics and personalized microbiome therapy (future of medicine)

With increasing discovery by science of the wide-ranging impact of the human microbiome on health, focused therapies such as probiotics and tailored microbiome therapy are beginning to appear as new medical tools for prevention as well as treatment.

1. Role of Probiotics

Probiotics refer to live bacteria that, in sufficient quantity, restore and maintain a balance of healthy microbials. Probiotics can be found naturally in fermented foods (such as yogurt, kefir, and kimchi) and also as supplements.

Benefits:

- Restore gut microbiota after the use of antibiotics
- Enhance digestion and relieve bloating
- Enhance immune function
- Provide relief for IBS symptoms as well as certain inflammatory bowel disorders
- May also affect mood and cognitive processes through the gut-brain axis

- Probiotics are not generic, however; various strains do different jobs and may have a different impact per person

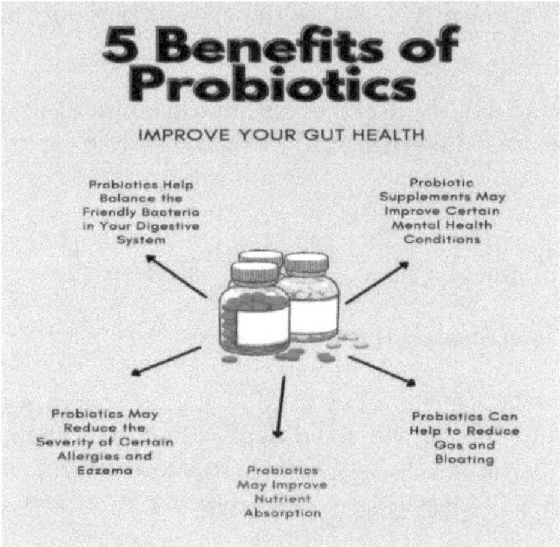

Figure 16: Shows benefits of probiotics

2. Personalized Microbiome Therapy

This new area of research hopes to individualize treatments according to an individual's specific microbial profile. Through testing different aspects of the gut microbiome, physicians can ascertain imbalances or deficiencies and then prescribe interventions ranging from diet changes and specific probiotics to fecal microbiota transplants (FMT) in extreme cases.

Potential Applications:

- Autoimmune and metabolic disease
- Neurological disorders (such as depression and autism)
- Cancer support therapy
- Precision nutrition and personalized medicine

A Glimpse into the Future

In the years to come, microbiome-based treatments and diagnoses could be common in medical practice. The method changes the approach from symptom treatment to cause healing, hence, making the treatments more individualized, natural, and prevent

References

Arrieta, M.-C., et al. (2015). The intestinal microbiome in early life: health and disease. Frontiers in Immunology, 6, Article 427. https://doi.org/10.3389/fimmu.2015.00427

Belkaid, Y., & Hand, T. W. (2014). Role of the microbiota in immunity and inflammation. Cell, 157(1), 121–141. https://doi.org/10.1016/j.cell.2014.03.011

Belkaid, Y., & Hand, T. W. (2014). Role of the microbiota in immunity and inflammation. Cell, 157(1), 121–141. https://doi.org/10.1016/j.cell.2014.03.011

Belkaid, Y., & Hand, T. W. (2014). Role of the microbiota in immunity and inflammation. Cell, 157(1), 121–141. https://doi.org/10.1016/j.cell.2014.03.011

Belkaid, Y., & Harrison, O. J. (2017). Homeostatic immunity and the microbiota. Immunity, 46(4), 562–576. https://doi.org/10.1016/j.immuni.2017.04.008

Bischoff, S. C., et al. (2014). Intestinal permeability – a new target for disease prevention and therapy. BMC Gastroenterology, 14, 189. https://doi.org/10.1186/s12876-014-0189-7

Blaser, M. J., & Falkow, S. (2009). What are the consequences of the disappearing human microbiota? Nature Reviews Microbiology, 7(12), 887–894. https://doi.org/10.1038/nrmicro2245

Bloomfield, S. F., et al. (2016). Time to abandon the hygiene hypothesis: New perspectives on allergic disease, the human microbiome, infectious disease prevention and the role of targeted hygiene. Perspectives in Public Health, 136(4), 213–224. https://doi.org/10.1177/1757913916650225.

Bokulich, N. A., Chung, J., Battaglia, T., et al. (2016). Antibiotics, birth mode, and diet shape microbiome maturation during early life. Science Translational Medicine, 8(343), 343ra82. https://doi.org/10.1126/scitranslmed.aad7121

Camilleri, M. (2019). Leaky gut: Mechanisms, measurement and clinical implications in humans. Gut, 68(8), 1516–1526. https://doi.org/10.1136/gutjnl-2019-318427

Carabotti, M., Scirocco, A., Maselli, M. A., & Severi, C. (2015). The gut-brain axis: Interactions between enteric microbiota, central and enteric nervous systems. Annals of Gastroenterology, 28(2), 203–209.

Carding, S., Verbeke, K., Vipond, D. T., Corfe, B. M., & Owen, L. J. (2015). Dysbiosis of the gut microbiota in disease. Microbial Ecology in Health and Disease, 26(1), 26191. https://doi.org/10.3402/mehd.v26.26191

Carding, S., Verbeke, K., Vipond, D. T., Corfe, B. M., & Owen, L. J. (2015). Dysbiosis of the gut microbiota in disease. Microbial Ecology in Health and Disease, 26(1), 26191. https://doi.org/10.3402/mehd.v26.26191

Clemente, J. C., Ursell, L. K., Parfrey, L. W., & Knight, R. (2012). The impact of the gut microbiota on human health: An integrative view. Cell, 148(6), 1258–1270. https://doi.org/10.1016/j.cell.2012.01.035

Cryan, J. F., & Dinan, T. G. (2012). Mind-altering microorganisms: The impact of the gut microbiota on brain and behaviour. Nature Reviews Neuroscience, 13(10), 701–712. https://doi.org/10.1038/nrn3346

Cryan, J. F., & Dinan, T. G. (2012). Mind-altering microorganisms: The impact of the gut microbiota on brain and behaviour. Nature Reviews Neuroscience, 13(10), 701–712. https://doi.org/10.1038/nrn3346

Cryan, J. F., & Dinan, T. G. (2012). Mind-altering microorganisms: The impact of the gut microbiota on brain and behaviour. Nature Reviews Neuroscience, 13(10), 701–712. https://doi.org/10.1038/nrn3346

Cryan, J. F., et al. (2019). The microbiota-gut-brain axis. Physiological Reviews, 99(4), 1877–2013. https://doi.org/10.1152/physrev.00018.2018

David, L. A., et al. (2014). Diet rapidly and reproducibly alters the human gut microbiome. Nature, 505(7484), 559–563. https://doi.org/10.1038/nature12820

Dinan, T. G., & Cryan, J. F. (2017). Gut instincts: microbiota as a key regulator of brain development, ageing and neurodegeneration. Journal of Physiology, 595(2), 489–503. https://doi.org/10.1113/JP273106

Dominguez-Bello, M. G., et al. (2010). Delivery mode shapes the acquisition and structure of the initial microbiota across multiple body habitats in newborns. PNAS, 107(26), 11971–11975. https://doi.org/10.1073/pnas.1002601107

Dominguez-Bello, M. G., et al. (2010). Delivery mode shapes the acquisition and structure of the initial microbiota across multiple body habitats in newborns. PNAS, 107(26), 11971–11975. https://doi.org/10.1073/pnas.1002601107

El-Salhy, M., Hatlebakk, J. G., Hausken, T. (2020). The role of fecal Microbiota transplantation in the treatment of irritable bowel syndrome. Therapeutic Advances in Gastroenterology, 13, 1–13. https://doi.org/10.1177/1756284820934586

Fasano, A. (2012). Leaky gut and autoimmune diseases. Clinical Reviews in Allergy Immunology, 42(1), 71–78. https://doi.org/10.1007/s12016-011-8291-x

Foster, J. A., & McVey Neufeld, K.-A. (2013). Gut–brain axis: How the microbiome influences anxiety and depression. Trends in Neurosciences, 36(5), 305–312. https://doi.org/10.1016/j.tins.2013.01.005

Foster, J. A., Rinaman, L., & Cryan, J. F. (2017). Stress & the gut-brain axis: Regulation by the microbiome. Neurobiology of Stress, 7, 124–136. https://doi.org/10.1016/j.ynstr.2017.03.001

Fung, T. C., Olson, C. A., & Hsiao, E. Y. (2017). Interactions between the microbiota, immune and nervous systems in health and disease. Nature Neuroscience, 20(2), 145–155. https://doi.org/10.1038/nn.4476

Gensollen, T., Iyer, S. S., Kasper, D. L., & Blumberg, R. S. (2016). How colonization by microbiota in early life shapes the immune system. Science, 352(6285), 539–544. https://doi.org/10.1126/science.aad9378

Gilbert, J. A., Blaser, M. J., Caporaso, J. G., et al. (2018). Current understanding of the human microbiome. Nature Medicine, 24(4), 392–400. https://doi.org/10.1038/nm.4517

Gilbert, J. A., Blaser, M. J., Caporaso, J. G., Jansson, J. K., Lynch, S. V., & Knight, R. (2018). Current understanding of the human microbiome. Nature Medicine, 24(4), 392–400. https://doi.org/10.1038/nm.4517

Hagan, T., Cortese, M., Rouphael, N., et al. (2019). Antibiotics-Driven Gut Microbiome Perturbation Alters Immunity to Vaccines in Humans. Cell, 178(6), 1313–1328.e13. https://doi.org/10.1016/j.cell.2019.08.010

Hill, C., Guarner, F., Reid, G., et al. (2014). Expert consensus document: The International Scientific Association for Probiotics and Prebiotics consensus statement on the scope and appropriate use of the term probiotic. Nature Reviews Gastroenterology & Hepatology, 11, 506–514. https://doi.org/10.1038/nrgastro.2014.66

Honda, K., & Littman, D. R. (2016). The microbiota in adaptive immune homeostasis and disease. Nature, 535(7610), 75–84. https://doi.org/10.1038/nature18848

Honda, K., & Littman, D. R. (2016). The microbiota in adaptive immune homeostasis and disease. Nature, 535(7610), 75–84. https://doi.org/10.1038/nature18848

Human Microbiome Project Consortium. (2012). Structure, function and diversity of the healthy human microbiome. Nature, 486(7402), 207–214. https://doi.org/10.1038/nature11234

Langdon, A., Crook, N., & Dantas, G. (2016). The effects of antibiotics on the microbiome throughout development and alternative approaches for therapeutic modulation. Genome Medicine, 8(1), 39. https://doi.org/10.1186/s13073-016-0294-z

Lozupone, C. A., Stombaugh, J. I., Gordon, J. I., Jansson, J. K., & Knight, R. (2012). Diversity, stability and resilience of the human gut microbiota. Nature, 489(7415), 220–230. https://doi.org/10.1038/nature11550

Lynch, S. V., & Pedersen, O. (2016). The human intestinal microbiome in health and disease. New England Journal of Medicine, 375(24), 2369–2379. https://doi.org/10.1056/NEJMra1600266

Lynn, D. J., Benson, S. C., Lynn, M. A., & Pulendran, B. (2022). Modulation of immune responses to vaccination by the microbiota: Implications and potential mechanisms. Nature Reviews Immunology, 22, 33–46. https://doi.org/10.1038/s41577-021-00609-2

Lynn, M. A., Tumes, D. J., Choo, J. M., et al. (2018). Early-life antibiotic-driven dysbiosis leads to dysregulated vaccine immune responses in mice. Cell Host & Microbe, 23(5), 653–660.e5. https://doi.org/10.1016/j.chom.2018.04.009

Marchesi, J. R., & Ravel, J. (2015). The vocabulary of microbiome research: A proposal. Microbiome, 3(1), 31. https://doi.org/10.1186/s40168-015-0094-5

Marchesi, J. R., et al. (2016). The gut microbiota and host health: A new clinical frontier. Nature Reviews Microbiology, 13(10), 673–690. https://doi.org/10.1038/nrmicro.2016.128

Marchesi, J. R., et al. (2016). The gut microbiota and host health: A new clinical frontier. Gut, 65(2), 330–339. https://doi.org/10.1136/gutjnl-2015-309990

Mayer, E. A., et al. (2015). Gut/brain axis and the microbiota. Journal of Clinical Investigation, 125(3), 926–938. https://doi.org/10.1172/JCI76304

Microbiology: an introduction / Gerard J. Tortora, Berdell R. Funke, Christine L. Case. 11th ed.

Microbiology: principles and explorations/Jacquelyn G. Black. 9th ed

Petersen, C., & Round, J. L. (2014). Defining dysbiosis and its influence on host immunity and disease. Cell Microbiology, 16(7), 1024–1033. https://doi.org/10.1111/cmi.12308

Rook, G. A. W. (2010). 99th Dahlem Conference on Infection, Inflammation and Chronic Inflammatory Disorders: Darwinian medicine and the 'hygiene' or 'old friends' hypothesis. Clinical & Experimental Immunology, 160(1), 70–79. https://doi.org/10.1111/j.1365-2249.2010.04133.x

Rooks, M. G., & Garrett, W. S. (2016). Gut microbiota, metabolites and host immunity. Nature Reviews Immunology, 16(6), 341–352. https://doi.org/10.1038/nri.2016.42

Sanders, M. E., Merenstein, D. J., Reid, G., Gibson, G. R., & Rastall, R. A. (2019). Probiotics and prebiotics in intestinal health and disease: From biology to the clinic. Nature Reviews Gastroenterology & Hepatology, 16(10), 605–616. https://doi.org/10.1038/s41575-019-0173-3

Sender, R., Fuchs, S., & Milo, R. (2016). Revised estimates for the number of human and bacteria cells in the body. PLoS Biology, 14(8), e1002533. https://doi.org/10.1371/journal.pbio.1002533

Sherwin, E., Bordenstein, S. R., Quinn, J. L., Dinan, T. G., & Cryan, J. F. (2019). Microbiota and the social brain. Science, 366(6465), eaar2016. https://doi.org/10.1126/science.aar2016

Singh, R. K., Chang, H. W., Yan, D., et al. (2017). Influence of diet on the gut microbiome and implications for human health. Journal of Translational Medicine, 15, Article 73. https://doi.org/10.1186/s12967-017-1175-y

Sonnenburg, E. D., & Sonnenburg, J. L. (2019). The ancestral and industrialized gut microbiota and implications for human health. Nature Reviews Microbiology, 17, 383–390. https://doi.org/10.1038/s41579-019-0191-8

Thursby, E., & Juge, N. (2017). Introduction to the human gut microbiota. Biochemical Journal, 474(11), 1823–1836. https://doi.org/10.1042/BCJ20160510

Vancamelbeke, M., & Vermeire, S. (2017). The intestinal barrier: A fundamental role in health and disease. Expert Review of Gastroenterology & Hepatology, 11(9), 821–834. https://doi.org/10.1080/17474124.2017.1343143

West, C. E., Renz, H., Jenmalm, M. C., et al. (2015). The gut microbiota and inflammatory noncommunicable diseases: Associations and potentials for gut microbiota therapies. Journal of Allergy and Clinical Immunology, 135(1), 3–13. https://doi.org/10.1016/j.jaci.2014.11.010

Yatsunenko, T., et al. (2012). Human gut microbiome viewed across age and geography. Nature, 486(7402), 222–227. https://doi.org/10.1038/nature11053

Zimmermann, P., & Curtis, N. (2018). The influence of the intestinal microbiome on vaccine responses. Vaccine, 36(30), 4433–4439. https://doi.org/10.1016/j.vaccine.2018.06.028

Zmora, N., Suez, J., & Elinav, E. (2019). You are what you eat: Diet, health and the gut microbiota. Nature Reviews Gastroenterology & Hepatology, 16(1), 35–56. https://doi.org/10.1038/s41575-018-0061-2

Zmora, N., Zeevi, D., Korem, T., et al. (2018). Personalized gut mucosal colonization resistance to empiric probiotics is associated with unique host and microbiome features. Cell, 174(6), 1388–1405.e21. https://doi.org/10.1016/j.cell.2018.08.041

ThankYou!

FSC
www.fsc.org
MIX
Papier | Fördert gute Waldnutzung
FSC® C083411

Kolibri 360 GmbH
Ferdinand-Jühlke-Straße 7
99095 Erfurt, Deutschland
produktsicherheit@kolibri360.de

Druck:
CPI Druckdienstleistungen GmbH
im Auftrag der
Kolibri 360 GmbH
Ferdinand-Jühlke-Str. 7
99095 Erfurt